AF228309

MILITARY MACHINES
DRONES
in Action

MARI BOLTE

Lerner Publications ◆ Minneapolis

Lerner Publications Company
An imprint of Lerner Publishing Group, Inc.
241 First Avenue North
Minneapolis, MN 55401 USA

For reading levels and more information, look up this title at www.lernerbooks.com.

Main body text set in Aptifer Sans LT Pro.
Typeface provided by Adobe Systems.

Library of Congress Cataloging-in-Publication Data

Names: Bolte, Mari, author.
Title: Drones in action / by Mari Bolte.
Description: Minneapolis : Lerner Publications , [2024] | Series: Military machines (Updog Books) | Includes bibliographical references and index. | Audience: Ages 8–11 | Audience: Grades 4–6 | Summary: "Explore the seas and skies with drones, some of the US military's most innovative machines. Discover the types of missions drones undertake and how they help soldiers stay one step ahead on the battlefield"— Provided by publisher.
Identifiers: LCCN 2022046085 (print) | LCCN 2022046086 (ebook) | ISBN 9781728491684 (library binding) | ISBN 9798765603383 (paperback) | ISBN 9781728498867 (ebook)
Subjects: LCSH: Drone aircraft—United States—Juvenile literature.
Classification: LCC UG1242.D7 B645 2023 (print) | LCC UG1242.D7 (ebook) | DDC 363.2028/4—dc23/eng/20220930

LC record available at https://lccn.loc.gov/2022046085
LC ebook record available at https://lccn.loc.gov/2022046086

Manufactured in the United States of America
2-1010669-51117-2/13/2024

Table of Contents

EYE IN THE SKY

A dark shape streaks through the sky. The drone flies silently over a group of people. It continues its mission unseen.

The drone scans a nearby mountain range. It takes a video and records the weather conditions.

The drone swoops around and heads back to its base.

No one is inside the drone. Its pilot sits in a
room miles away.

A sensor operator helps by keeping track of the drone's cameras and weapons.

UP NEXT! KEEP AN EYE OUT.

DRONES IN THE DISTANCE

Aerial drones can fly into dangerous areas without risk to a human pilot. The US Navy also has drone ships. People call them the "Ghost Fleet" because there are no sailors on board.

The first US military drones were built during
World War I (1914–1918).

They have only gotten faster and more powerful since then.

Drones are fast. Some can go five times the speed of sound.

The US military is working on a drone that may be able to fly more than 4,000 miles (6,437 km) per hour.

Some drones can only fly for short periods.

Others have larger batteries that allow them to fly for more than twenty-four hours without recharging.

Engine
AF
05 141

Wings
Camera
Landing Gear

Military drones have high-tech cameras. They watch the sky and the ground for danger.

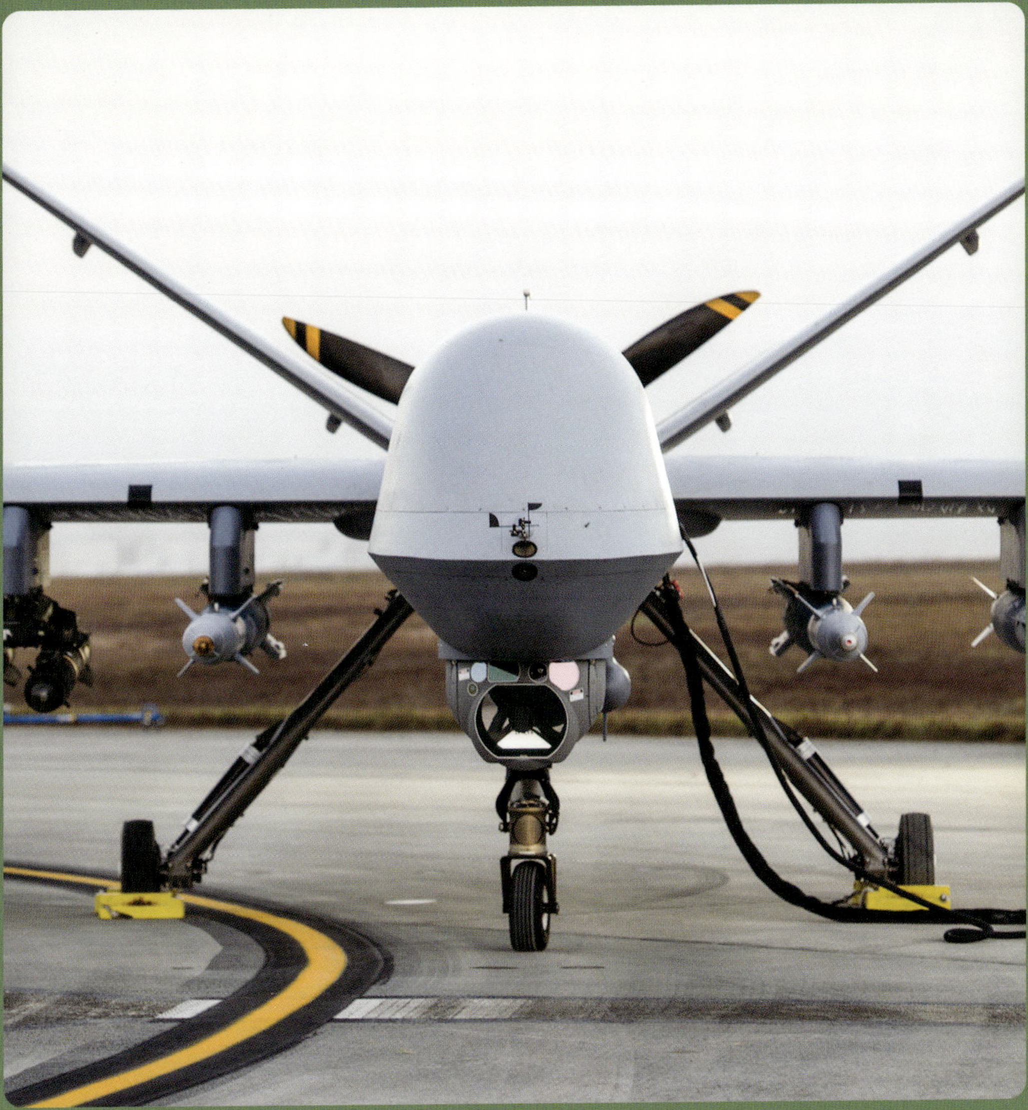

They can search for people and objects. Some drones also carry weapons.

UP NEXT!
TIME TO LAUNCH.

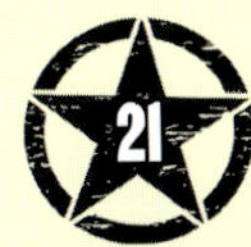

ALL SHAPES AND SIZES

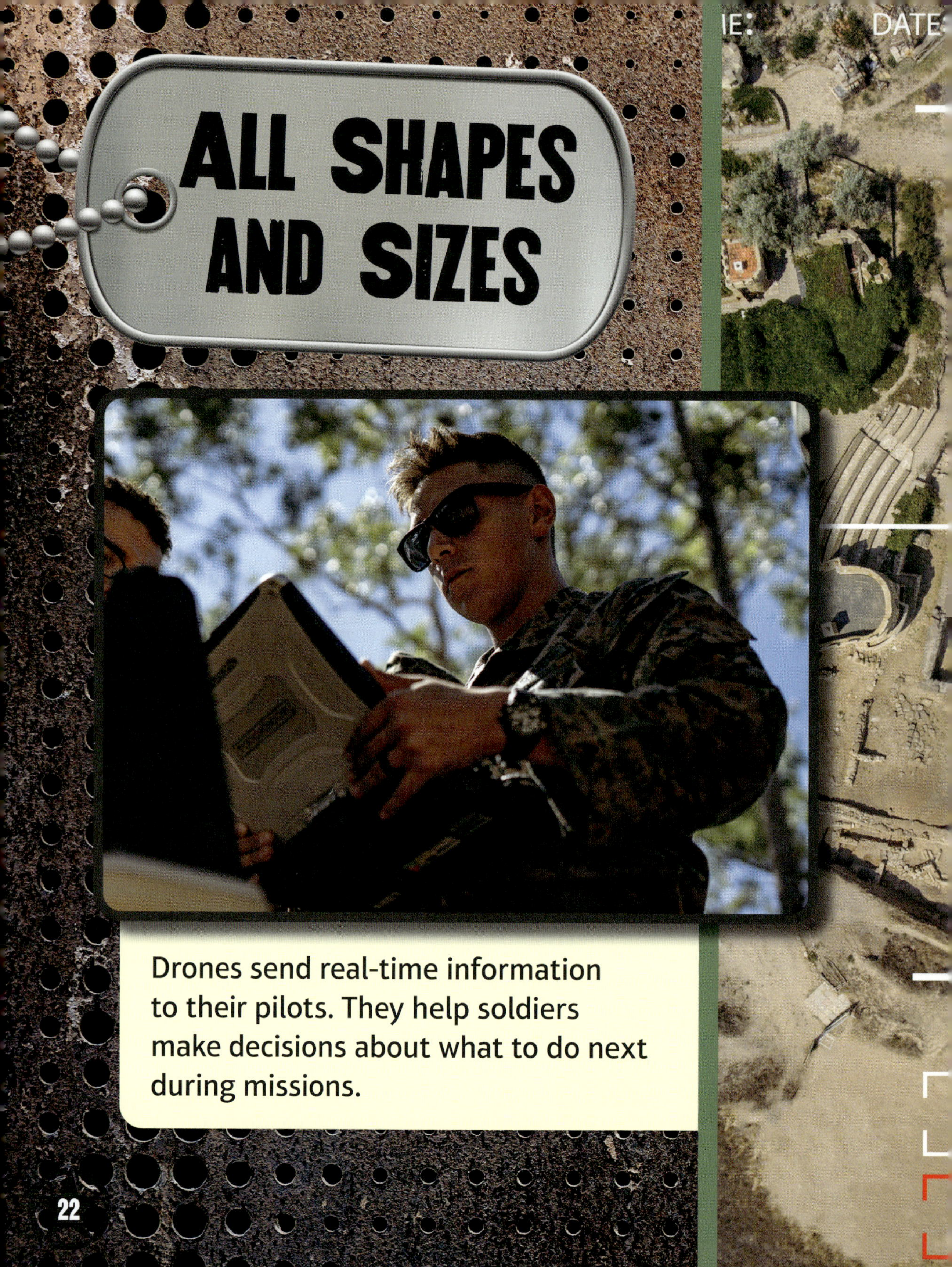

Drones send real-time information to their pilots. They help soldiers make decisions about what to do next during missions.

ALTITUDE: 1297m
FlIGHT SPEED: 1400km
300
100
16
20mm
30mm
00mm
23

Raven drones are small. They weigh 4.4 pounds
(2 kg) and are 4.5 feet (1.4 m) wide.

A soldier can launch a Raven by hand.

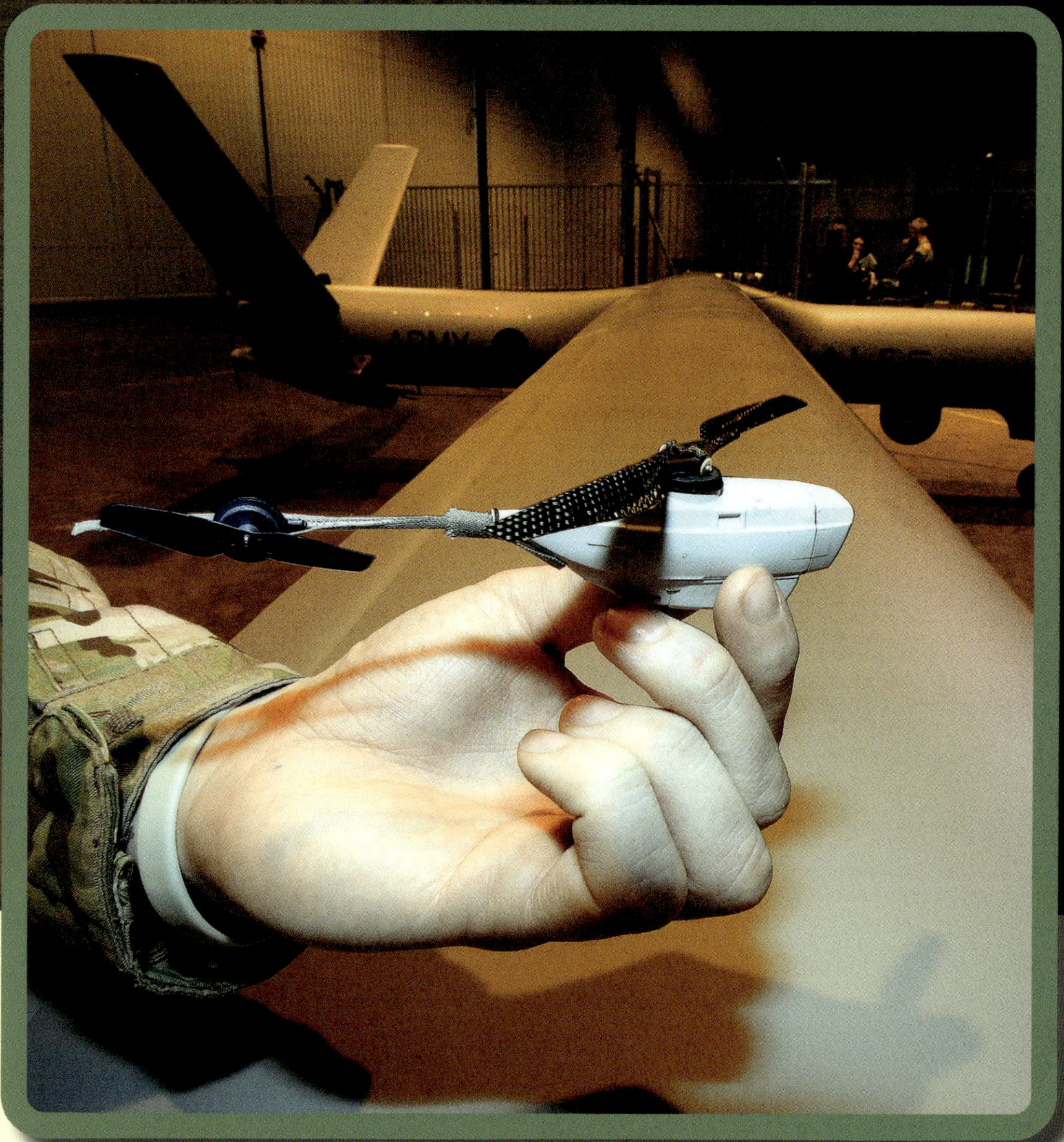

Black Hornets are even smaller than Ravens.
They fit in the palm of a hand.

They can fly for twenty-five minutes before needing to recharge.

The *Sea Hunter* is a drone ship. Its job is to look for submarines.

With so many drones ready for action, there will always be an eye on the skies and seas.

Glossary

aerial: happening in the air

base: a place used by the military to store equipment and train soldiers

sensor: a device that detects or measures something

speed of sound: the rate at which sound moves, which is around 760 miles (1,223 km) per second

submarine: a ship designed to operate underwater

Check It Out!

BuiltIn: Drone Technology: What Is a Drone?
https://builtin.com/drones

Chandler, Matt. *Drones*. Minneapolis: Bellwether Media, 2022.

Drone Rush: Military Drones – The New Air Force
https://www.dronerush.com/military-drones-air-force-navy-marines-cia-10853/

McKinney, Donna Bowen. *Inventing Drones*. Lake Elmo, MN: Focus Readers, 2022.

Schaefer, Lola M. *Flying Robots*. Minneapolis: Lerner Publications, 2021.

Unmanned Aerial Vehicle Facts for Kids
https://kids.kiddle.co/Unmanned_aerial_vehicle

Index

Photo Acknowledgments

Image credits: gremlin/Getty Images, p.4; JoeLena/Getty Images, p.5; Standard store88/Shutterstock, p.6; Sgt. William Parsons/DVIDS, p.7; Ethan Miller/Staff/Getty Images, p.8; Ethan Miller/Staff/Getty Images, p.9; Moraima Johnston/DVIDS, p.10; Petty Officer 1st Class Tyler Fraser/DVIDS, p.11; U.S. Air Force/National Museum of the United States Air Force, p.12; Stocktrek Images/Getty Images, p.13; Alan Radecki/DVIDS, p.14; Hermeus/flickr, p.15; Becki Bryant/DVIDS, p.16; Courtesy Photo/DVIDS, p.17; Everett Collection/Shutterstock, p.18; 1st Lt. Daniel de La Fe/DVIDS, p.20; Ilka Cole/DVIDS, p.21; Cpl. Lydia Gordon/DVIDS, p.22; WindVector/Shutterstock, p.23; Maj. Will Cox/DVIDS, p.24; Spc. Ryan Lucas/DVIDS, p.25; Nigel Roddis/Stringer/Getty Images, p.26; 1st Sgt. lekendrick stallworth/DVIDS, p.27; Petty Officer 2nd Class Aiko Bongolan/DVIDS, p.28; Staff Sgt. Thomas Calvert/DVIDS, p.29

Design element: SEAN GLADWELL/Getty Images; geengraphy/Getty Images

Cover: HIGH-G Productions/Stocktrek Images/Getty Images